W9-ALM-477

Everyday Science Experiments in the Kitchen

John Daniel Hartzog

The Rosen Publishing Group's
PowerKids Press™
New York

Some of the experiments in this book are designed for a child to do together with an adult.

Published in 2000 by The Rosen Publishing Group, Inc.
29 East 21st Street, New York, NY 10010

Photo Illustrations by Shalhevet Moshe

First Edition

Book Design: Michael de Guzman

Hartzog, John Daniel
    Everyday science experiments in the kitchen / by John Daniel Hartzog.
        p.   cm. — (Science surprises)
    Summary: Provides experiments that explore scientific phenomena occuring in the kitchen.
    ISBN 0-8239-5456-0 (alk. paper)
    1. Food—Experiments—Juvenile literature. 2. Science—Experiments—Juvenile literature. 3. Cookery—Juvenile literature. [1. Food—Experiments. 2. Science—Experiments. 3. Experiments.] I. Title. II. Series: Hartzog, John Daniel. Science surprises.
    TX355.H294   1999
    507.8—dc21
                                                                                   99-13876
                                                                                         CIP

Manufactured in the United States of America

Contents

Science in the Kitchen

Strange things happen in the kitchen. Water magically appears from the faucet then disappears down the drain. The refrigerator stays cold on the hottest days of the year. Simple ingredients like milk, flour, and eggs are gooey when mixed together, but once that mixture bakes in the oven it changes into bread or cake.

The kitchen is usually a place for grown-ups. You don't have to be old, though, to understand how things work. Science can help you understand what goes on in the kitchen.

Science helps people to describe and understand the world. You can be a scientist. All you need is a puzzle or a question and an **experiment** to help you figure it out.

◀ *Sometimes kids need help understanding what happens in the kitchen.*

Dirty Copper

Do you ever have to wash the dishes? Did you know that lemon juice cleans better than soap and water? Well, at least on pennies. A dirty penny cannot be cleaned with ordinary soap and water. Pennies are made of a metal called **copper** (like some cooking pots). In a **chemical reaction**, copper combines with **oxygen** in the air to make copper oxide. This reaction makes the penny darker and duller-looking.

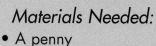

Materials Needed:
- A penny
- A glass
- Lemon juice

The lemon juice will make the pennies bright and shiny again. ▶

You can make the penny bright and shiny again. Squeeze some lemon juice into a glass. (If you do not have any lemons, you can use vinegar instead.) Leave the penny in the lemon juice for 10 minutes. The **acid** in the juice will **dissolve** the copper oxide. The penny will come out shiny, like new. You have just seen a chemical reaction.

Materials Needed:
- A glass
- 30 pennies
- A sponge

Water Holds Onto Water

Have you ever poured so much water into a glass that it spilled over the top? Here's an experiment that shows how water can rise above the top of the glass. Water is made of tiny **particles** called molecules. The molecules in water attract each other and try to hold on to one another. You can see this for yourself with this experiment.

Fill a small glass right to the top. Now drop the pennies in one at a time. Watch the water rise. What happens to the surface of the water? It should rise just a little over the top of the glass. When water touches air, the water forms a flexible skin to hold the water together. When water molecules hold onto other water molecules it's called surface tension.

◀ *Have a sponge ready in case some water spills.*

Ooblech

Do you know the difference between solids and liquids? Water and milk are liquids. A table, a pan, and a spoon are solids. Can you see and feel the difference between

Materials Needed:
- One bowl
- Five tablespoons of cornstarch
- Four tablespoons of water
- Food coloring (optional)

solids and liquids? Ooblech is a mixture that sometimes feels like a solid and sometimes feels like a liquid. Here's a recipe.

Mix the cornstarch, water, and a few drops of food coloring in a bowl. The ooblech will be thick and hard to stir. Now try to poke the mixture with a fork. Wow, can you believe how hard that is? Now push the fork into the mixture slowly. How did it get so soft? The ooblech seems hard when you poke it and soft when you push the fork gently into it. That's because the cornstarch does not dissolve in the water. When you poke the ooblech, the water is pushed out, so the ooblech seems solid. When you push it slowly, the water stays in and the grains of cornstarch move around more easily. Then ooblech acts like a liquid.

◀ *Stirring the ooblech.*

The final product. ▶

Blow Up a Balloon Without Even Blowing

With a household chemical reaction, you can make a soda bottle blow up a balloon. Find an empty soda bottle. With a funnel, pour the vinegar into the bottle. Cut a paper towel in half. Cut one half of the paper towel in half again so you have 1/4 of a paper towel. Pour two teaspoons of baking soda into the center of the paper towel. Roll the paper towel into a tube small enough to fit into the bottle. Drop the

Materials Needed:
- One soda bottle
- One funnel
- One paper towel
- Two teaspoons of baking soda
- A quarter cup of vinegar
- One balloon

Step 1 ▶

◀ Step 2

paper towel tube filled with baking soda into the bottle and quickly stretch the balloon over the neck of the bottle. The vinegar and baking soda work together to create a gas called carbon dioxide. The gas escapes from the vinegar in tiny bubbles that begin to fill the bottle. The bottle is already full of air so the new gas pushes the air into the balloon. The air begins to blow up the balloon.

◀ Step 3

Step 4 ▶

Kitchen Smells

Materials Needed:
- One balloon
- One funnel
- Vanilla extract
- One shoe box

Don't you love when you wake up in your bedroom and can smell breakfast cooking in the kitchen? Here is a neat experiment that might help you to understand how the smells get there from the kitchen.

Using the funnel, put a few drops of vanilla extract inside your balloon. Blow up the balloon and tie it off. Place the balloon inside a shoe box and close the lid. Come back in a few hours and open the lid. Somehow the vanilla smell fills the box, even though you put it inside the balloon. Smells are made of tiny particles that float in air. Smells spread out evenly

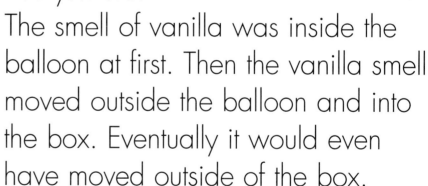

through osmosis. Osmosis means that if there is a lot of one smell in a certain place, it will spread into other places until there is an even amount of that smell everywhere.

The smell of vanilla was inside the balloon at first. Then the vanilla smell moved outside the balloon and into the box. Eventually it would even have moved outside of the box.

◀ Step 2

15

Place the celery in the water.

▼

Materials Needed:
- A glass
- Food coloring
- A stalk of celery
  (with leaves still attached)

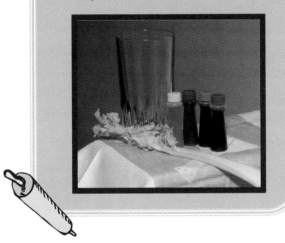

You can see how the water travels up the stalk. ▶

A Celery Straw

Have you ever noticed how celery **stalks** have all those lines running through them? Plants need water to stay alive. Those lines are **veins**. They let water move through the celery while it is growing. Water moves through the celery like a straw. Let's try an experiment that will show this happening.

Mix a drop of food coloring into a glass of water. Find a fresh piece of celery that still has some leaves on top. Put the celery in the water with the leaves pointing up. Place the glass and the celery next to a window. Check on your experiment the next day. Can you see the color moving up inside the stalk? The water is moving up the stalk through **transpiration**.

Cleaning Dirty Water

Materials Needed:
- Two jars
- One paper towel
- Dirt
- Water

The drinking water that comes out of your faucet comes from rivers, lakes, or **reservoirs**. Before it comes to you it has been cleaned. Drinking dirty water is not healthy for people. Try this experiment to find one way you can clean water at home.

Fill one jar with a few spoonfuls of dirt and then add some water. Stir the water until it looks muddy. Place this jar on a stack of two or three books. Place an empty jar on the table so that it is below the first jar. Starting at a corner, roll a paper towel into a tight, long tube. Put one end of the paper towel

◀ *Step 1*

Step 2 ▶

into the dirty water. Put the other end into the lower, empty jar. As you watch you will see the paper towel getting wet. The water travels through the paper towel because it clings to the surface of the paper towel. The dirt will not travel through the paper towel. This process is called **filtering**. This experiment filtered a jar of dirty water, just the way water from reservoirs is filtered before we can drink it.

Step 3 ▶

Chemical Changes

The everyday science of the kitchen often depends on chemistry. Chemistry is a science that helps us to understand the ingredients of things. Two important

Materials Needed:
- Boiled red cabbage
- Vinegar
- Baking soda
- Lemon juice
- Aspirin (crushed with the back of a spoon)
- Milk

ingredients in the kitchen are acids and **bases**. Acids and bases are hydrogen-based solutions that will make salt when they are mixed together. An acid is sour, like lemon juice or vinegar. A base is bitter, like aspirin.

To see the difference between acids and bases for yourself, boil some red cabbage (ask for some help from an adult).

The strained liquid from the cabbage will change colors when it is mixed with different things. In separate bowls, mix some cabbage juice first with some vinegar, then baking soda, then lemon juice, then  aspirin, and then milk. The cabbage will turn blue in a base and pink in an acid. You can tell whether each ingredient is an acid or a base by what color the cabbage juice turns. It's chemistry right in your own kitchen.

What Is Science?

Now you are a scientist. The kitchen is your laboratory. You saw water travel up a celery stalk, a balloon blow up on its own, and chemicals change colors right before your eyes. You have used your senses to answer questions and uncover mysteries. Scientists use careful eyes, ears, and noses to put clues together to solve a puzzle. You did not just guess your answers. You created experiments to test ideas and to answer questions. You mixed ingredients and watched the changes. Some of the changes were exciting because you made discoveries. Understanding science can help you to describe and see the world in new ways.

Glossary

acid (A-sid) A hydrogen-based solution that reacts with bases to form salt. Acids usually taste sour.

base (BAYS) A hydrogen-based solution that reacts with acids to form salt. Bases usually taste bitter.

chemical reaction (KEM-ih-kul ree-AK-shun) A change in the chemical nature of two substances.

copper (KOP-er) A reddish-brown metal that is a chemical element. It is a soft and inexpensive metal used to make pennies.

dissolve (dih-ZOLV) To seem to disappear when mixed with a liquid.

experiments (ehks-PER-ih-ment) A test or tests that are used to find out something or to test the effect of something.

filtering (FIL-ter-ing) When water passes through a cloth to remove dirt.

oxygen (AHKS-ih-jen) A gas without color, taste, or odor that plants and animals breathe in and need to survive.

particle (PAR-tih-kul) A small piece of something.

reservoir (REZ-ur-vwar) A place where water is collected and stored for use.

stalk (STAHLK) A stem; the part of the celery we eat.

transpiration (tran-spu-RAY-shun) The process by which plants give off water into the atmosphere.

vein (VAYN) A rib or channel in a plant.

Index

Web Sites:

You can learn more about kitchen science on the Internet.
Check out this Web site: http://freeweb.pdq.net/headstrong